Copyright © Lucien Khan, 2013

The right of Lucien Khan to be identified as the author of this work has been asserted by him in accordance with sections 77 and 78 of the Copyright, Designs and Patents Act, 1988.

ISBN-13: 978-1492162032

ISBN-10: 1492162035

First Published August 2013

The 216 Letter Hidden Name Of God

For my daughter Erin. Life is filled with glorious purpose and great meaning. You were all the proof I ever needed.

Chapters

Chapter 1

The 216 Letter Sacred Hidden Name of God .... 11 – 12

Chapter 2

The Fibonacci 60 Digit Repeat Cycle.. 14 – 23

Chapter 3

The Seed Of Life.. 25 – 27

Chapter 4

Metatron's Cube - Divine Revelation... 29 – 32

Chapter 5

The Shemhamphorasch (The Divided Name of God).................................... 34 – 38

Chapter 6

Super Intelligence (The Signature of God)... 40 – 41

Chapter 7

The Sacred Numbers 54 & 108 & 216.. 43 – 48

Chapter 8

Angels and Angles.. 50 – 55

CHAPTER 1

The 216 Letter Sacred Hidden Name of God

In the Ancient Kabala it was believed that there is a secret 216 Letter Hidden Name of God and that once this name is revealed it will usher in the Messianic age. The age of global peace and harmony.

I have found the Secret 216 letter name of God hidden inside Metatron's Cube and the Seed Of Life.

Metatron was an archangel and is sometimes said to be Enoch, a pious, good man, who ascended up into heaven to sit at God's side and is thought to be one of only a few of the angels who is allowed to look upon the countenance of God.

This next excerpt is from the Encyclopaedia Mythica:

> "Sitting next to God, Enoch was instructed in wisdom, and using his skills as a scribe, prepared three hundred and sixty-six books. When he learned everything, a most significant thing happened. God revealed to him great secrets - some of which are even kept secret from the angels!
>
> These included the secrets of Creation, the duration of time the world will survive, and what will happen after its demise. At the end of these discussions, Enoch returned to earth for a limited time, to instruct everyone, including his sons, in all he learned. After thirty days, the angels returned him to Heaven.
>
> And then the divine transformation took place. Additional wisdom and spiritual qualities caused Enoch's height and breadth to become equal to the height and breadth of the earth.

> God attached thirty-six wings to his body, and gave him three hundred and sixty-five eyes, each as bright as the sun.
>
> His body turned into celestial fire -- flesh, veins, bones, hair, all metamorphosed to glorious flame. Sparks emanated from him, and storms, whirlwind, and thunder encircled his form. The angels dressed him in magnificent garments, including a crown, and arranged his throne. A heavenly herald proclaimed that from then on his name would no longer be Enoch, but Metatron, and that all angels must obey him, as second only to God."

This discovery or REVELATION I have made will UNITE the world and FREE all people. I call upon the Enlightened to contemplate this discovery and to share it with all mankind.

As you read this, always bear in mind that that in Ancient times they used to practice Gematria, Gematria was a traditional Jewish system of assigning numerical value to a word or phrase.

In essence Numbers were used as Letters or Words.

The 216 Numbers I found reveal Gods Hidden and Sacred name. They prove that God is real and the Universe is an Intelligent Creation and not a random Singularity event as per the Big Bang.

I must declare upfront that I myself do not follow any single religion. I simply believe that 'GOD IS EVERYTHING'.

The relevance of the discovery presented in this book are my own views based upon my understanding of the world and my interpretation of religious doctrines.

I can guarantee you though, that no Intelligent person, be that a Mathematician, Physicist, Theologian or scholar of any worth will be able to look you in the eye and say that what I have found is merely a random coincidence.

This is truly a remarkable discovery as you will no doubt be able to ascertain for yourself.

THE FIBONACCI SEQUENCE

AND

THE 60 DIGIT REPEAT CYCLE

CHAPTER 2

The Fibonacci 60 Digit Repeat Cycle

If you look at the numbers in the Fibonacci Sequence you will find that the last digit in each number forms part of a pattern that repeats after every 60th number and this 60 number pattern repeats all the way into infinity.

What is the Fibonacci Sequence?

The Fibonacci Sequence is a series of numbers:

0, 1, 1, 2, 3, 5, 8, 13, 21, 34, ...

The next number is found by adding up the two numbers before it. So the next number in the above example would be: 55.

I have listed the first 72 numbers in the sequence and highlighted the last number in red. Notice that after the first 60 numbers (0-59) the last number starts to repeat. This 60 number pattern repeats all the way into infinity.

0 : 0
1 : 1
2 : 1
3 : 2
4 : 3
5 : 5
6 : 8
7 : 13
8 : 21

9 : 34
10 : 55
11 : 89
12 : 144
13 : 233
14 : 377
15 : 610
16 : 987
17 : 1597
18 : 2584
19 : 4181
20 : 6765
21 : 10946
22 : 17711
23 : 28657
24 : 46368
25 : 75025
26 : 121393
27 : 196418
28 : 317811
29 : 514229
30 : 832040
31 : 1346269
32 : 2178309
33 : 3524578
34 : 5702887
35 : 9227465
36 : 14930352
37 : 24157817
38 : 39088169
39 : 63245986
40 : 102334155
41 : 165580141

42 : 267914296
43 : 433494437
44 : 701408733
45 : 1134903170
46 : 1836311903
47 : 2971215073
48 : 4807526976
49 : 7778742049
50 : 12586269025
51 : 20365011074
52 : 32951280099
53 : 53316291173
54 : 86267571272
55 : 139583862445
56 : 225851433717
57 : 365435296162
58 : 591286729879
59 : 956722026041
60 : 1548008755920
61 : 2504730781961
62 : 4052735537881
63 : 6557470319842
64 : 10610209857723
65 : 17167680177565
66 : 27777890035288
67 : 44945570212853
68 : 72723460248141
69 : 117669030460994
70 : 190392490709135
71 : 308061521170129
72 : 498454011879264

SPECIAL NOTE: THE 216TH NUMBER IN THE FIBONACCI SEQUENCE IS –

216: 619220451666590135228675387863297874269396512

And if you add up all the individual digits in that number it adds up to 216. (Coincidence?)

The 60 numbers in the Fibonacci Sequence repeat cycle are:

0, 1, 1, 2, 3, 5, 8, 3, 1, 4, 5, 9, 4, 3, 7, 0, 7, 7, 4, 1, 5, 6, 1, 7, 8, 5, 3, 8, 1, 9, 0, 9, 9, 8, 7, 5, 2, 7, 9, 6, 5, 1, 6, 7, 3, 0, 3, 3, 6, 9, 5, 4, 9, 3, 2, 5, 7, 2, 9, 1

This was all known.

You should take the time to go online and Google the Fibonacci Sequence. There are many online reference pages listing the Fibonacci Sequence. Computers are constantly probing deeper and deeper into the Sequence looking for patterns.

So what's the big deal?

Like I said, this was all known. What I did, that was totally unique is, I arranged those 60 digits from the repeat cycle around the circumference of a circle (as seen in **Diagram 1** on next page) and I found the results so amazing and revealing that they are simply Divine.

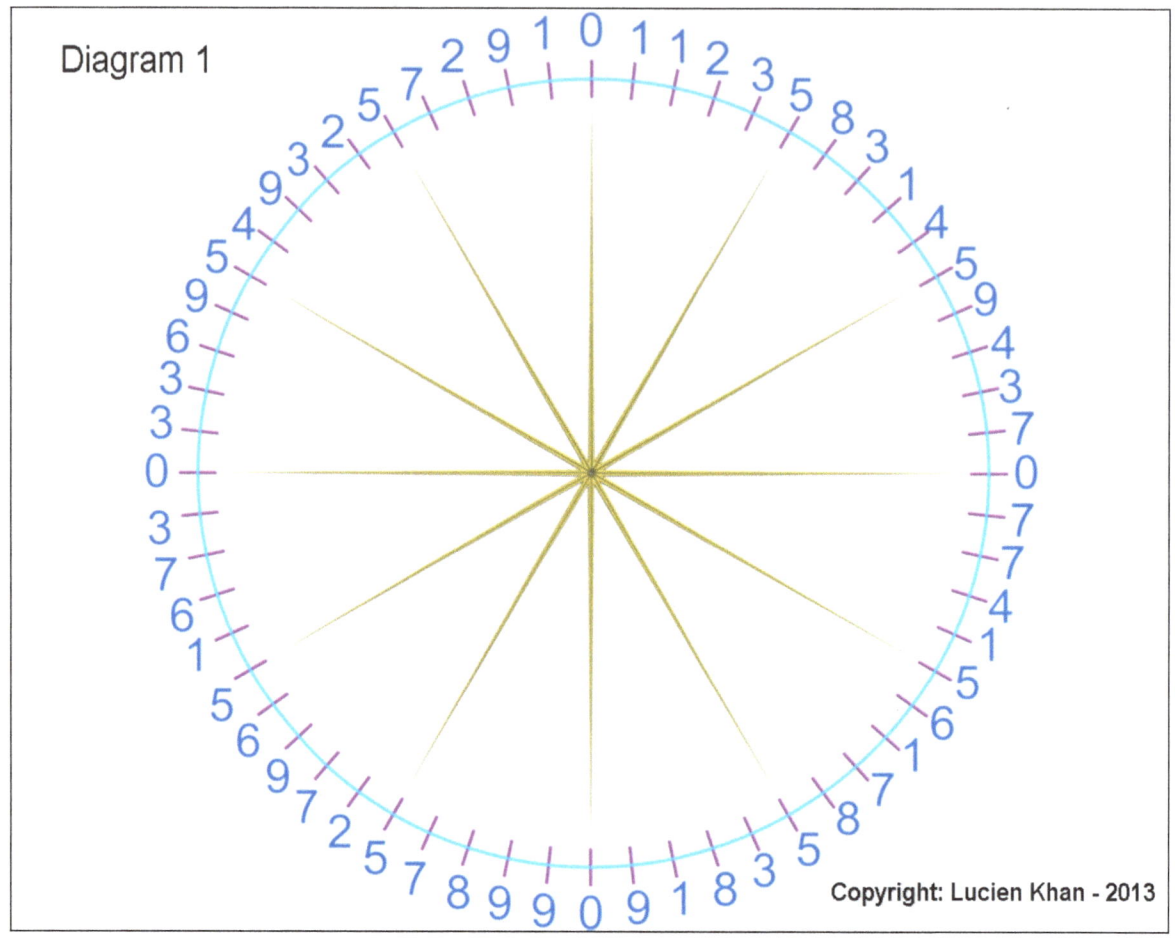

Looking at **Diagram 1** the first thing you will notice is that the 'Zero's' from the 60 Digit repeat sequence align perfectly with the 4 cardinal points on a compass. And if you look at the number diagonally adjacent or directly opposite each number, as shown by the star, you will see that they all add up to a perfect 10.

It is important to note that this 60 digit repeat cycle only exists because we use base 10 mathematics. This, as you will see, is vital to the way in which we view the entire universe.

BUT THERE IS SO MUCH MORE:

At this point I need to tell you why I am so fascinated by this 60 digit repeat pattern.

I have been working on a cosmological theory that states that the universe was not created in a singularity event as per the 'Big Bang' but rather that the universe is an eternal cycle of energy to mass and that this cycle repeats after every 60 billion years or rather after every 60 units of time. A "billion years" has no real external value and is simply our way of recording time using base 10 mathematics.

The number 6 is what is of ultimate importance and is crucial in my findings as 6 is the first whole entity in the physical universe. Every physical object must have 6 dimensions from its point of origin, being: **UP, DOWN, LEFT, RIGHT, BACKWARD & FORWARD**.

We get to 60 because we are using base 10 mathematics i.e:

6 x 10 = 60

6 squared or 6 x 6 = 36

36 x 10 = 360

The physical universe manifests in all 360 degrees.

6 cubed or 6 x 6 x 6 = 216

There are incredible predictions related to the number 216.

For instance: It is believed that the secret or hidden name of God contains 216 characters.

Now, bearing my cosmological theory in mind you can see why I was so astounded to find this 60 digit repeat cycle hidden in the Fibonacci Sequence.

But this could simply be a random pattern of 60 numbers right?

Yes, they could be. So I need to do more to qualify my theories and arguments.

If 6 x 6 x 6 = 216 which has divine connotations. What happens if I take 3 of those Fibonacci Circles and combine them as in **Diagram 2**?

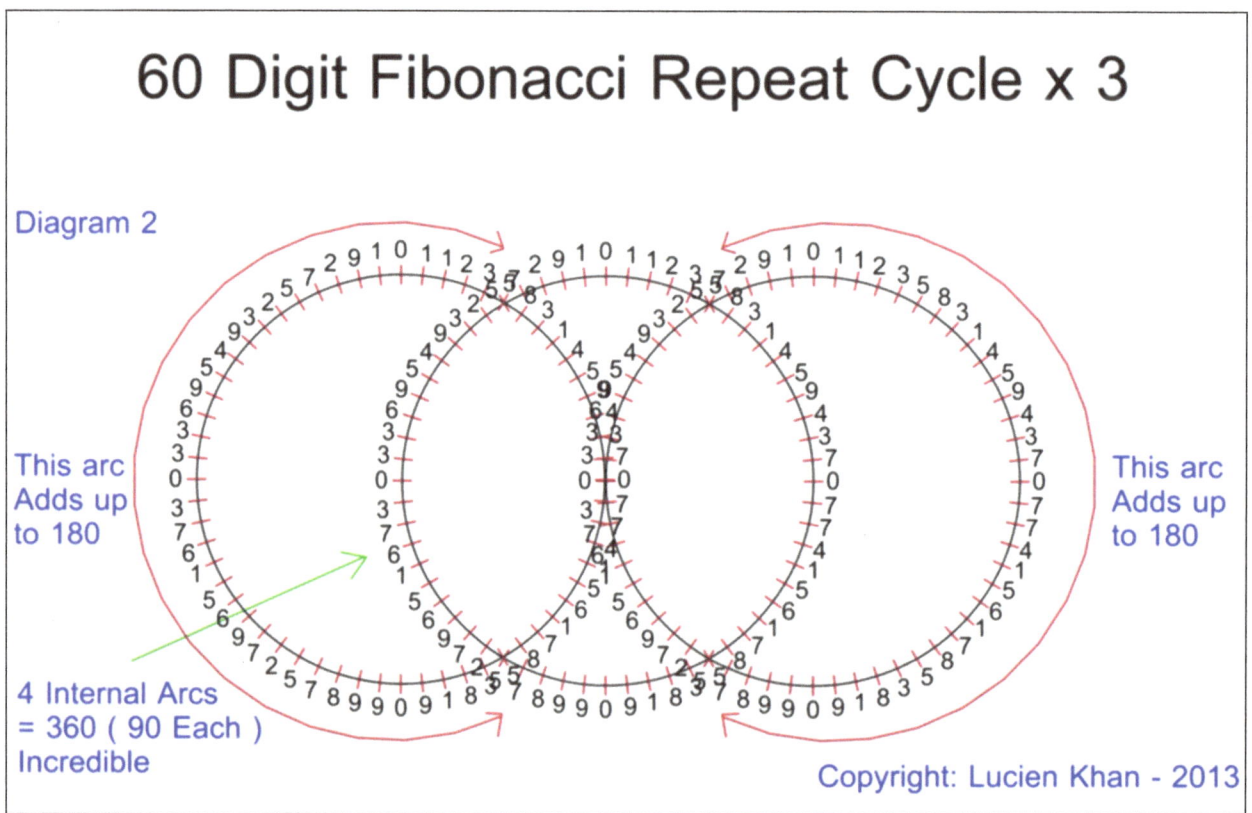

Clearly, these are not simply 60 random numbers repeating to all infinity. When 3 of these circles are aligned at the dead centre, as in Diagram 2, the circumference of the circles touch perfectly at their midpoints (Number 5 in the cycle) and I find that the 4 internal arcs add up to 360 and the 2 exterior arcs each add up to 180. (180 x 2 =360)

Those '5' point markers are located at exactly 120 degrees of arc.

So now I have a 60 digit repeat cycle that produces the number 360 when aligned in this order.

Remember my theory says that the universe recycles every 60 units of time and manifests in all 360 degrees.

Here are 60 numbers that produce the sum of 360 when aligned in this 3 circle pattern. Coincidence? You could still get away with saying that now, but wait till you see what else is hidden.

So what else is there?

If I align 2 circles (2 x 60 digit repeat cycles) with one on top of the other intersecting at the first or top number "5" markers as seen in **Diagram 3** below. The remaining numbers add up to 216.

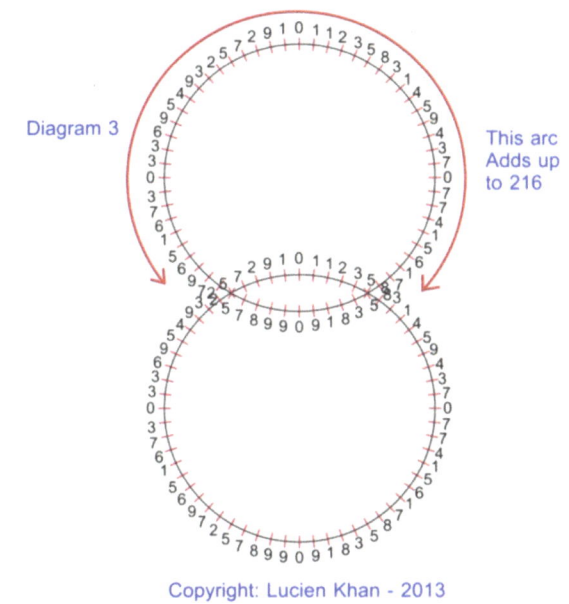

Now I have the Divine Number 216 in the Ascending. I think even the most sceptical will have to stop and take a closer look at all of this now.

This arc that adds up to exactly 216 can be seen in many depictions of saints or angels and is used to signify their Devine nature. Can it really be simple coincidence?

COULD ALL STILL BE COINCIDENCE RIGHT?

You won't be saying that at the end of this book. I guarantee it.

Do you know that mathematicians and scholars don't know precisely why we use 60 seconds and 60 minutes or 360 degrees. The answer is buried somewhere in antiquity and it was assumed that these were simply arbitrary units that we adopted for convenience sake.

It is thought that our ancient ancestors believed that the earth took 360 days to orbit the sun in 1 year and so they used 360 degrees for the circumference of a circle. Modern scholars simply adopted the 360 degrees.

To me, my discovery is proof that the first whole unit in the physical world is 6. Made up of the 6 dimensions mentioned earlier being **UP, DOWN, LEFT, RIGHT, BACKWARD & FORWARD.**

Every object in the physical universe manifests from its point of origin and extends in these 6 dimensions.

Traditional physicists say we have only 3 dimensions being XYZ or length, width and height. But I beg to differ. That is an egotistical view of the world based on our point of view and not from the point of origin or centre of the object being observed.

An object has the traditional XYZ plus –XYZ making 6 dimensions.

Every object extends from its centre or finite point of origin outward.

6 is the first 'whole' 6 x 10 using base 10 mathematics gives us 60 units.

6 x 60 = 360. The universe manifests in all 360 degrees.

6 x 6 x 6 = 216 The Divine Trinity.

NB.

- 216 is the smallest untouchable number which is also a cube

- Plato's number: Plato alludes to the fact that 216 is equal to 6 x 6 x 6, where 6 is one of the numbers representing marriage since it is the product of the female 2 and the male 3. Plato was also aware of the fact that the sum of the cubes of the 3-4-5 Pythagorean triple is equal to 216. (3 cubed = 27, 4 cubed = 64 and 5 cubed = 125. Therefore, 27 + 64 + 125 = 216.)

- The total number of corner-angle degrees on the surface of a cube is 2160.

- The diameter of the moon is 2 159,14062 miles, you can round that off to 2160 statute miles.

- The years of a Zodiac Age are also 2160.

- All life is Carbon based and Carbon has 6 electrons, 6 protons and 6 neutrons.

THE SEED OF LIFE

CHAPTER 3

The Seed of Life

The Kabala teaches that God created the world with seven spiritual building blocks—His seven "emotional" attributes.

The Bible says that God created the world in 6 days and rested on the seventh. Here again we have reference to the number 6.

It is my belief that what was being revealed here is that God created the world in 6 dimensions. But we also see the number 7 being of great significance, both here and in the Islamic and many other cultures.

So far, I have only mentioned 6 dimensions.

But I knew that physicists counted 3 dimensions, what they call length, width and breadth, plus TIME. Time is described as the 4th dimension.

I use the same 3 dimensions (XYZ) but I also have (-xyz) which brings me to the six dimensions mentioned above, what I call **UP, DOWN, LEFT, RIGHT, BACKWARD & FORWARD** for want of better labels.

But I also know that time is the 7th dimension.

This number 7 is significant in all major religions that make reference to the 7 days or 7 building blocks.

So I aligned 7 of my Fibonacci 60 Digit Repeat Cycles as seen in **Diagram 4** (Next Page).

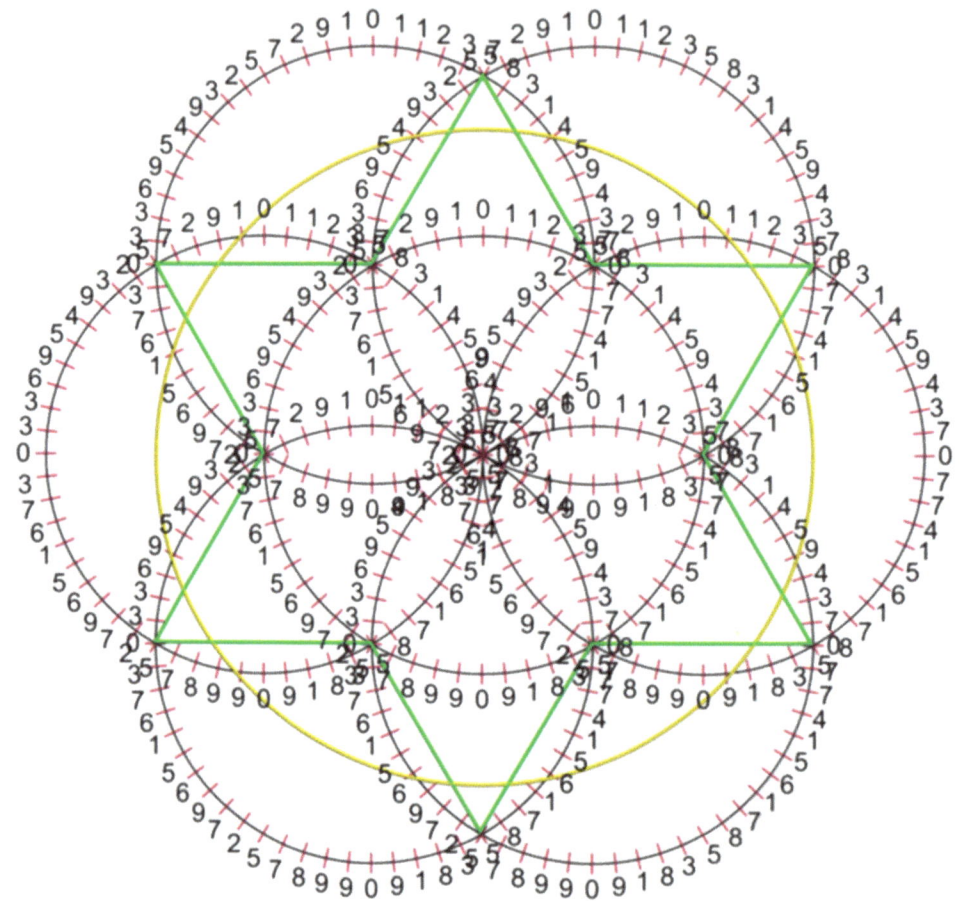

These 7 circles now form a famous pattern known as **"The Seed Of Life"**. The Seed of Life is a Divine Symbol or Sacred Geometry that appears throughout the ancient world as well as in works by Leonardo Da Vinci.

The way that the circles align perfectly on the number '5 & zero' makers is so amazing it can no longer be called "coincidence". Well, maybe by the stoically sceptical or cynical, but even they will be moved by what comes later.

When I calculate the points where the flower pattern intersect I get a perfect 60. Draw a circle anywhere inside that pattern (as indicated by the yellow circle) and the corresponding numbers on the 12 points where the circle intersects with the 'flower' pattern on the circumference adds up to 60.

This is no coincidence.

But there is so much more. When I superimpose the Star of David into the design (Green Star in **Diagram 4**) the 6 inside points add up to 60.

The 6 outer points add up to 40. Together they give me a perfect 100. This is a perfect 100% or whole using the "base-ten harmonic".

Believe it or not this is still only the tip of the ice-berg.

METATRON'S CUBE

DIVINE REVELATION

CHAPTER 4

Metatron's Cube Divine Revelation

This next part is nothing short of Divine and points to an Ancient history rich with mystery and hidden knowledge.

Metatron, as mentioned earlier, was a scribe of God and was said to have been entrusted with divine knowledge with regards to the secrets of Time and Eternal Life.

The **Diagram 5** (below) is a partial representation of Metatron's Cube which is considered to be a Sacred Diagram or Divine Geometry that contains all the secrets of the universe.

If you research Metatron's Cube you will find that it is a famous geometric figure and revered because it is said to contain all of the 5 platonic solids, being: The Tetrahedron, The Cube, The Octahedron, the Dodecahedron and The Icosahedron.

The totally amazing and incredible part is that when I create Metatron's Cube using my 60 digit repeat cycle I get the **Sacred Numbers 216 & 108** (Google these numbers). These are very important numbers in many religions and were considered sacred by Plato as well as Fibonacci.

108 x 2 = 216.

The secret name of God is said to contain 216 characters.

The diameter of the moon is 2160 miles

The solar radius is approximately 432,450 miles (2160 x 2 = 4320)

And the speed of light is approximately 432 squared or 186 624 miles per second (approximately).

60 Digit Fibonacci Repeat Cycle inside Metatron's Cube

The total sum of all the numbers in this diagram adds up to 3640 - but take the points where the star of David intersect as '5' you get 3640 - 40 = 3600

Diagram 5

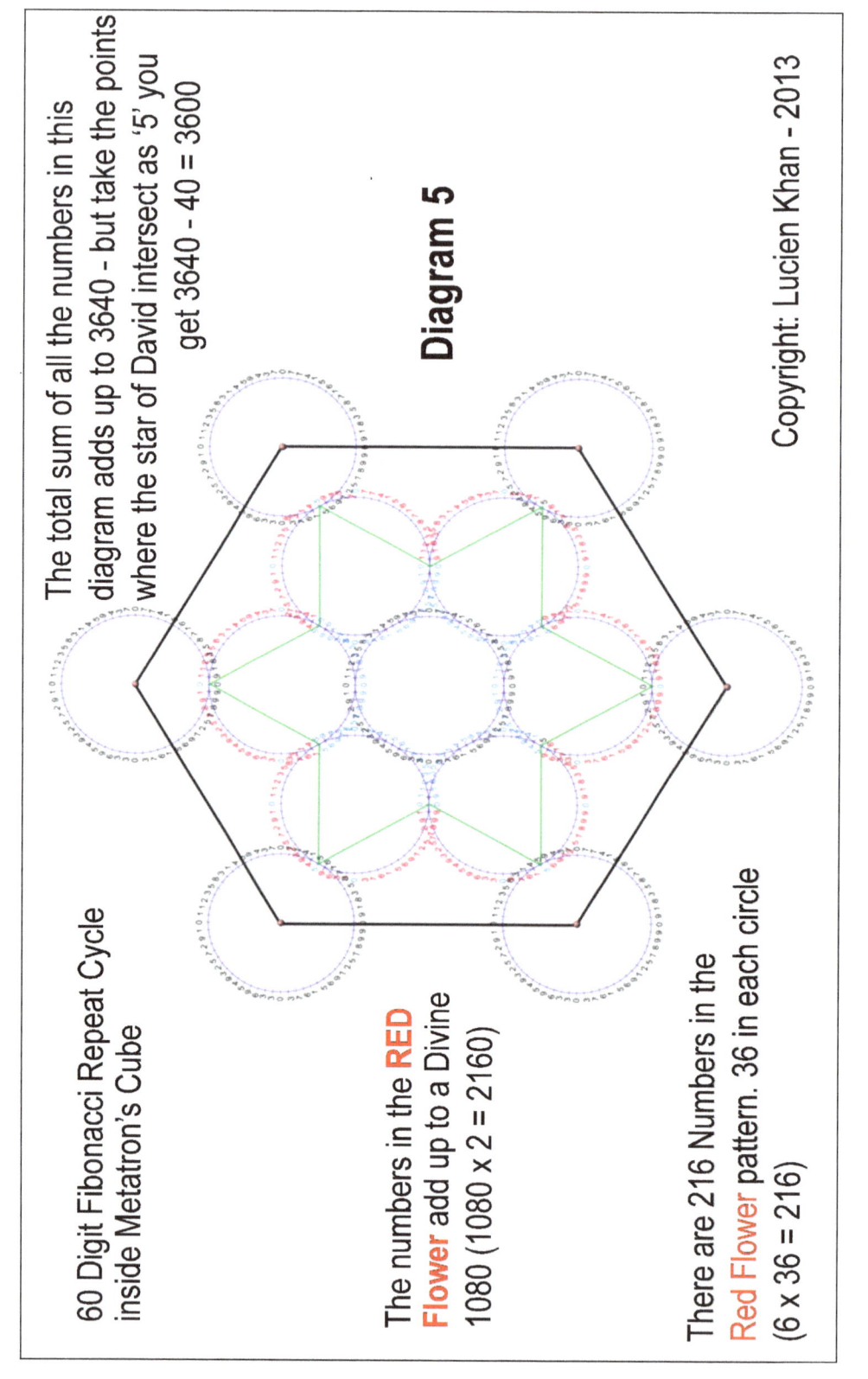

The numbers in the **RED Flower** add up to a Divine 1080 (1080 x 2 = 2160)

There are 216 Numbers in the Red Flower pattern. 36 in each circle
(6 x 36 = 216)

Copyright: Lucien Khan - 2013

The sum of the 60 numbers in each circle adds up to 280. There are 13 circles in Metatron's Cube.

Therefore in Total we have 280 x 13 = 3640

Take all those points where the Star Of David intersects not as (5 + 5) but as '5' and we get:

3640 – 40 (There are 8 points hence 5 x 8 = 40)

= 3600

3600 is a base 10 harmonic of the number 360, i.e 360 x 10 = 3600

This design is "PERFECT"

Bear in mind that Metatron's Cube is a Fractal

A Fractal is a curve or geometric figure, each part of which has the same statistical character as the whole.

To me the answer is as clear as day. The universe recycles after every 60 units of time and manifests in 360 degrees. This is how the Universe exists for all eternity. Metatron's Cube is a Fractal of the "WHOLE" and is clearly showing us why we use units of 60 and 360.

But not only is Metatron's Cube generating the number 360 it is also generating the Divine numbers "108" and "216". You need to research (Goggle) these numbers and learn about their historic and religious significance.

The sum of the numbers highlighted in the Flower pattern in Red is a Divine 1080 (That's 108 x 10). There are 216 numbers in that pattern. Each circle contributes 36 numbers. 6 x 36 = 216.

Why was it believed that the Secrets of the Universe were Hidden inside Metatron's Cube and Sacred Geometry? And why was it believed that the Secret Name of God contained 216 letters?

We didn't know that the end numbers in the Fibonacci Sequence recycle after every 60 digits into infinity until we had computers.

According to the Code of Carl Munck, many Ancient Sites like the Great Pyramid, Stonehenge, The Circle Fort in Ohio, The Golo Circle in Germany, Cuicuilco in Mexico are all encoded with the number 216.

Now, although I maintain that you should not naively believe everything you read. I do encourage you to not cynically dismiss everything as superstitious nonsense simply because there are so many predictions that don't come to pass. The world is filled with great mystery. If Carl Munck is right, even partially, then what were the Ancients trying to tell us or what had they been told regarding the 216 number.

NB: Apologies for the poor resolution on the images as I try to render the entire cube so that you may see the patterns clearly.

You can find higher resolution images online as well as Youtube video's that I created explaining all of this in more detail, where I zoom into the images for a closer look.

To find my online material simply search Youtube for "Lucien Khan and Metatron's Cube".

I cannot begin to explain the sheer "INTELLIGENCE' involved here, this is undoubtedly a big step toward proving once and for all that the universe is not a random singularity event but it is an Intelligent Creation and recycles for all eternity.

Anyone still sceptical? There's more.

THE SHEMHAMPHORASCH

(THE DIVIDED NAME OF GOD)

CHAPTER 5

The Shemhamphorasch (The Divided Name of God)

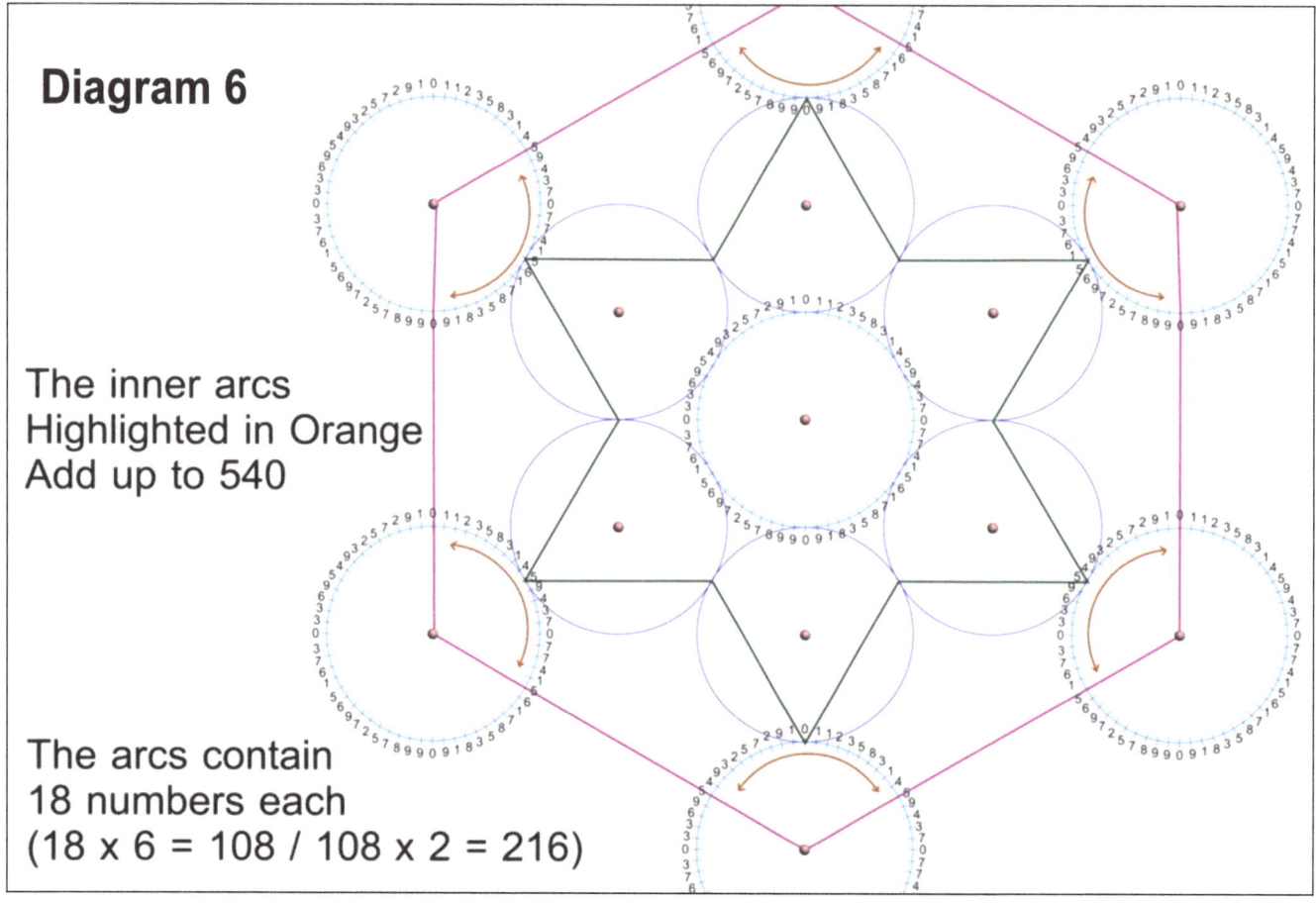

Diagram 6

The inner arcs Highlighted in Orange Add up to 540

The arcs contain 18 numbers each (18 x 6 = 108 / 108 x 2 = 216)

Bearing **Diagram 6** (Above) in mind look at this next table showing the Divided Name of God. Each section contains 54 letters. 54 X 4 = 216.

SHEMHAMPHORASCH: THE DIVIDED NAME (TABLE COPIED FROM WIKIPEDIA)

	18	17	16	15	14	13	12	11	10	9	8	7	6	5	4	3	2	1
י	כK	לL	הH	הH	מM	יI	הH	לL	אA	הH	כK	אA	לL	מM	עO	סS	יI	וV
	לL	אA	קQ	רR	בB	זZ	הH	אA	לL	זZ	הH	כK	לL	הH	לL	יI	לL	הH
	יI	וV	מM	יI	הH	לL	עO	וV	דD	יI	תTh	אA	הH	שSh	מM	טT	יI	וV

	36	35	34	33	32	31	30	29	28	27	26	25	24	23	22	21	20	19
ה	מM	כK	לL	יI	וV	לL	אA	רR	שSh	יI	הH	נN	חCh	מM	יI	נN	פP	לL
	נN	וV	הH	חCh	שSh	כK	וV	יI	אA	רR	אA	תTh	הH	לL	יI	לL	הH	וV
	דD	קQ	חCh	וV	רR	בB	מM	יI	הH	תTh	אA	הH	וV	הH	יI	כK	לL	וV

	54	53	52	51	50	49	48	47	46	45	44	43	42	41	40	39	38	37
ו	נN	נN	עO	הH	דD	וV	מM	עO	עO	סS	יI	וV	מM	הH	יI	רR	חCh	אA
	יI	נN	מM	חCh	נN	הH	יI	שSh	רR	אA	לL	וV	יI	הH	יI	הH	עO	נN
	תTh	אA	מM	שSh	יI	וV	הH	לL	יI	לL	הH	לL	כK	הH	זZ	עO	מM	יI

	72	71	70	69	68	67	66	65	64	63	62	61	60	59	58	57	56	55
ה	מM	הH	יI	רR	חCh	אA	מM	דD	מM	עO	יI	וV	מM	הH	יI	נN	פP	מM
	וV	יI	בB	אA	בB	יI	נN	מM	חCh	נN	הH	מM	צTz	רR	יI	מM	וV	בB
	מM	יI	מM	הH	וV	עO	קQ	בB	יI	וV	הH	בB	רR	חCh	לL	מM	יI	הH

The Shemhamphorasch is a corruption of the Hebrew term Shem ha-Mephorash meaning "The explicit Name of God".

I must reemphasize that I myself do not follow any single religion. I am merely fascinated by the fact that the ancient's believed that the secret name of God contained 216 Characters and that when this name was revealed it would usher in the Messianic Age.

I believe that God is everything. All my findings indicate that all religions had a pure origin or source that was trying to educate or direct us.

I need to explain to you what exactly was driving me on and directing me to keep searching for the answer. Every time I faltered I would discover something new, like this table divided into 18 x 12 parts. So I kept delving deeper and deeper.

I will tell you more about my personal journey later. Let me first show you just how complex and Super Intelligent this all is.

This next image is a Christian Diagram and this one shows the 72 Names of God. 72 x 3 = 216.

Once again we see a connection throughout the ancient world related to these sacred numbers.

It may interest you to know that in each cycle in my rendering of Metatron's Cube there are 60 numbers, so in total, for this entire rendering of Metatron's Cube using the 60 digit cycle there are 60 x 13 = 780 numbers. This is the total amount of physical numbers and not the sum of all the numbers added together which gave us the 3600 I spoke of earlier. (See **Diagram 5**)

But each circle has 4 zero's so subtract 4 x 13 which is 52 you get 780 – 52 = 728.

And, remember the "5's" intersect at 8 points (See Diagram 5). So we only count those numbers once. Now we have 728 – 8 = 720. And 720 is a base 10 harmonic of 72.

The more I study Metatron's Cube using the 60 digit Fibonacci repeat cycle, the more convinced I am that the ancient's were far more knowledgeable or informed than we ever thought.

They were told about the 72 Names of Angels. They were told to study Metatron's Cube which is based on the Tree of Life and they were told to table the 216 letter name into 18 columns x 12 rows. Why?

Clearly we don't use 60 minutes and 360 degrees because it was a silly superstition or because it was convenient. No, we use these units because they point to a Perfect Universe created by a Supreme Intelligence that was always trying to guide and educate mankind.

There are still many secrets hidden in this design. Secrets that I am convinced will unlock the gates to the Messianic Age.

I didn't pick 60 random numbers. The 60 numbers I am using come directly from the Fibonacci Sequence. The same sequence that generates the Golden Ratio and the Golden Angle.

I have done a great deal of research into the Golden Ratio and the Golden Angle."

Physicists and Cosmologists calculate the age of the universe to be roughly 13.75 Billion years old. I tie this number directly to the Golden Angle which is 137.5 degrees. (13.75 x 10).

We are at a unique point in time where new things occur in nature. I believe we are living in a time of revelations.

Take another look at how perfectly the 60 repeating numbers from the Fibonacci Sequence fit into the face of a clock. The 5's and 0's align perfectly with the 12 hours on a clock face. The 4 zero's also clearly mark out the 4 cardinal points on a compass.

Do we simply use 60 minutes on a clock because our Ancestors did? Or is there a greater far more intelligent reason?

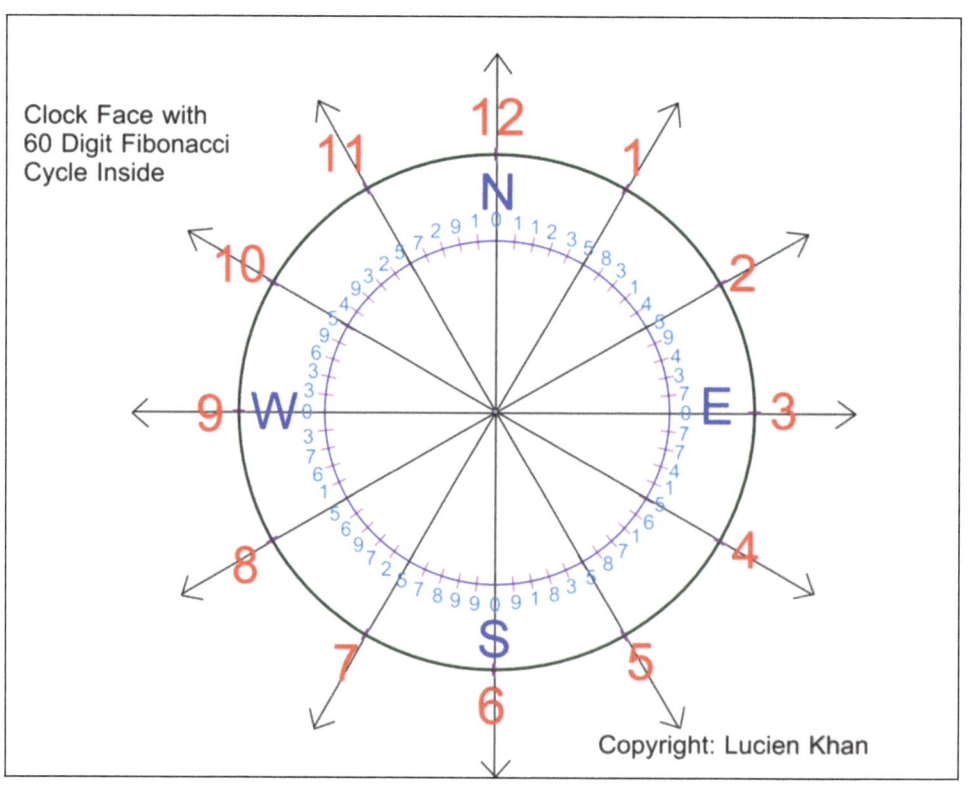

SUPER INTELLIGENCE
(THE SIGNATURE OF GOD)

CHAPTER 6

Super Intelligence (The Signature of God)

Metatron's Cube is like a blueprint for the universe, it shows us how the universe is created using fractal geometry and it contains the hidden 216 letter Sacred Signature of God.

Now look at Diagram 7

From here we can see how the number pattern grows in increments of '6' and spreads out into all 360 degrees. 6 x 360 = 2160. (1080 x 2 = 2160).

The green flower pattern is an exact copy of the numbers in the red flower pattern and will flower into infinity.

Not only do these 2 Divine Numbers appear here inside Metatron's Cube but they are also reflected in every aspect of our Physical 6 dimensional Universe.

The size of the moon is 2160 Miles. The solar radius is approximately 432,450 miles. (216 x 2 = 432).

The speed of light is approximately 186,282 miles per second (432 squared = 186,624).

There are twelve Astrological Ages in total; one for each constellation of the zodiac. Each Age lasts for approximately 2160 years.

According to the code of Carl Munck many Ancient sites like Stonehenge encode the numbers 60, 360 and 2160.

You can clearly see how everything is a fractal and builds up from Metatron's Cube and natures first pattern, 'The Seed of Life".

This is why the Ancients were told to study Metatron's Cube and why the Seed of Life came to be revered and obsessed over. The secrets to the Universe were hidden inside Sacred Geometry all this time.

Metatron's Cube constructed using
Fibonacci 60 Digit Repeat Cycle

Diagram 7
Copyright: Lucien Khan

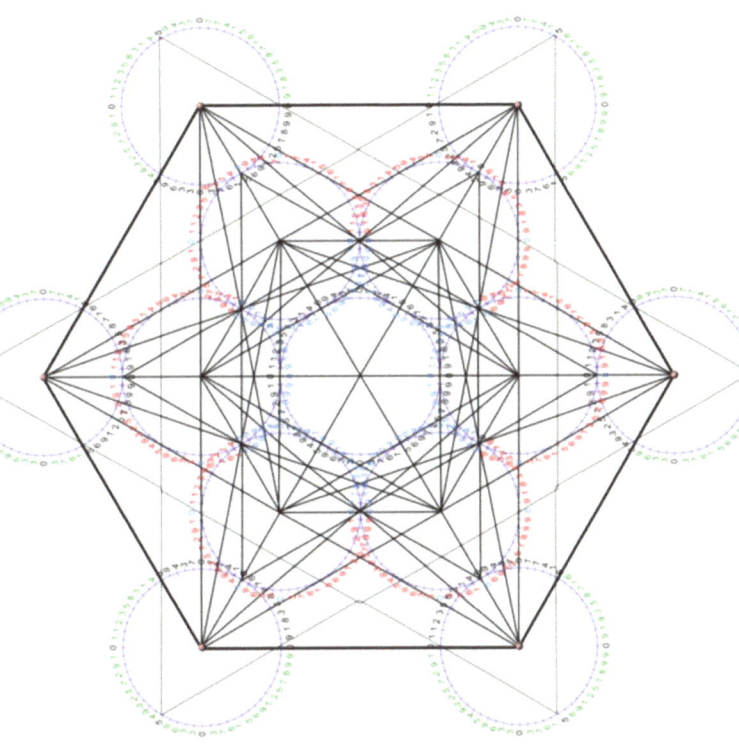

The numbers in the red inner flower add up to the Divine 1080.

There are 216 numbers in the red flower pattern. (Being 6 x 36)

The numbers 108 & 216 have great religious significance and in the Cabala it is said that the Secret Name of God has 216 letters.

The numbers in the green outter flower add up to the Divine 1080.

There are 216 numbers in the green flower. (Being 6 x 36)

Here you can see how this pattern will now repeat into infinity.

THE SACRED NUMBERS: 54, 108 AND 216

CHAPTER 7

The Sacred Numbers 54 & 108 & 216.

Really Study this next Diagram 8:

DIAGRAM 8

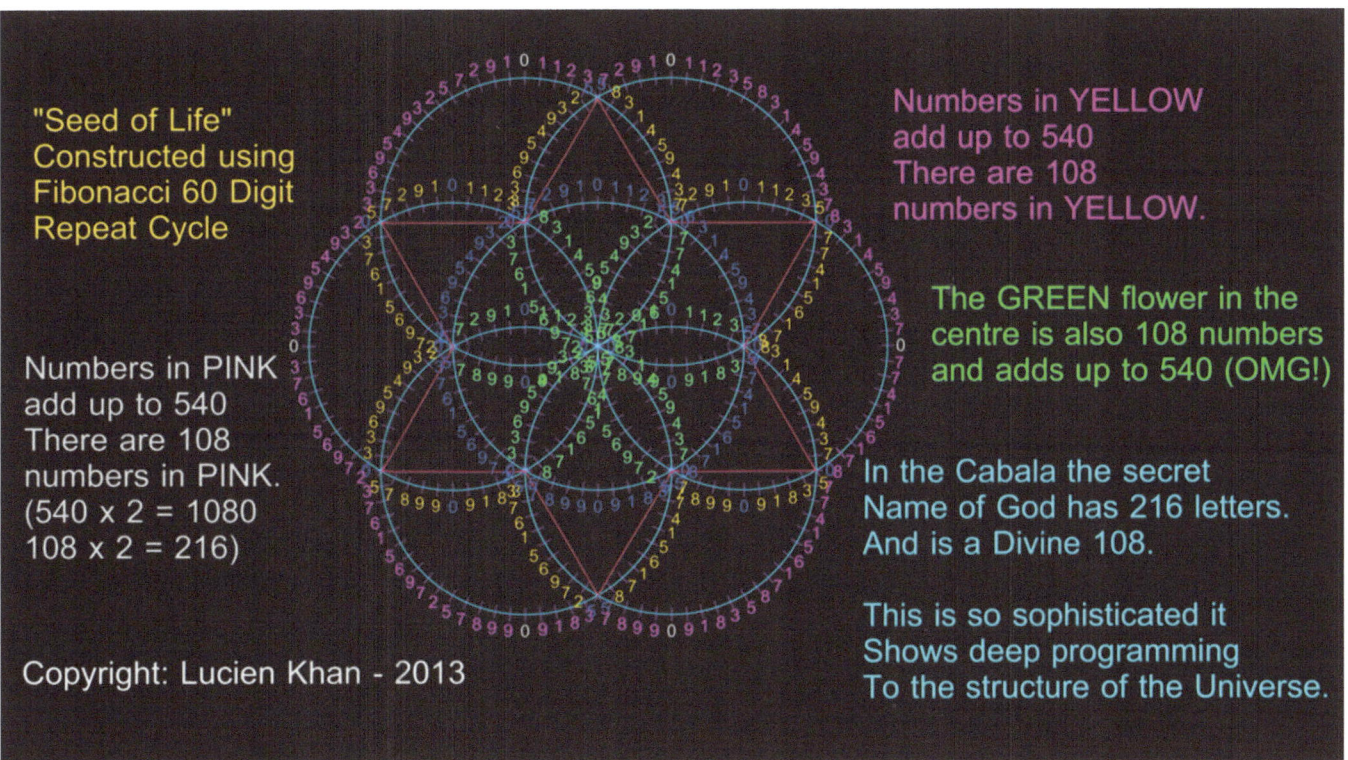

Like I said at the beginning of this book. There is no Intelligent person alive who could look at this diagram and say that this is all a coincidence.

If you have ever played Sudoku or tried to create a magic number box then you will be able to understand the sheer brilliance at work here.

The Green Flower pattern in the middle consists of 108 numbers and the sum is 540.

That Flower then grows into the Orange (Yellow) Flower pattern which consist of 108 numbers and the sum is 540.

Finally it blooms into the Pink Flower pattern which also consist of 108 numbers and adds up to 540.

This is no coincidence. This shows deep programming and Super Intelligent Design to the very structure of all Geometry.

Can you even comprehend what it would take to get 60 numbers to do this? And then for this entire Flower or Seed of Life to bloom into Metatron's Cube and once again reflect the 108 and 216 number pattern, it is INCREDIBLE.

The sheer art and beauty of this is beyond us. Even if the numbers 54, 108 and 216 had no special meaning, just to get numbers to do this, to follow this pattern is 'Uber' Intelligence.

But, throw into the equation thousands of years of history teaching us that these numbers are sacred. Wow. How can anyone tell you now that there is no God? If they did, then you would be the fool for believing them.

The number 108 comes up in a variety of contexts. There are 54 letters in the Sanskrit alphabet. Each has a masculine/feminine or Shiva/Shakti aspect: 54 x 2 is 108.

On the sacred geometrical configuration known as the Sri Yantra, there are points called Marmas where three lines intersect, and there are 54 such intersections. Each intersection also has masculine/feminine or Shiva/Shakti qualities.

In the human body, Marmas or Marmastanas are energy intersections similar to chakras. There are said to be 108 Marmas in the subtle body. Also, it is said that there are a total of 108 main energy lines or Nadis that radiate from the heart chakra.

Thus, the number 108 relates to the Sri Yantra as well as the human body. (Until now nobody could say why???)

The diameter of the sun is 108 times the diameter of the Earth.

In the Krishna tradition, there were said to be 108 Gopis or maid-servants of Krishna.

There are 108 forms of traditional Indian dance.

In the Vedas there are 108 Upanishads as listed in the Mundakopanishad.

There is said to be 108 Tantric texts.

Sikh: The Sikh tradition has a mala of 108 knots tied in a string of wool, rather than beads.

Buddhism: Some Buddhists carve 108 small Buddhas on a walnut for good luck. Some ring a bell 108 times to celebrate a new year. There are said to be 108 virtues to cultivate and 108 defilements to avoid.

Chinese: The Chinese Buddhists and Taoists use a 108 bead mala, which is called Su-Chu, and has three dividing beads, so the mala is divided into three parts of 36 each. Chinese astrology says that there are 108 sacred stars.

Stages of the soul: It is said that Atman, the human soul or centre goes through 108 stages on the journey.

First man in space: The first manned space flight lasted 1 Hour and 48 minutes, and was on April 12, 1961 by Yuri Gagarin, a Soviet cosmonaut.

1 hour and 48 minutes is 108 minutes.

When the number 25,920 is divided into the 12 signs of the Zodiac we get 2,160.

In the book of Revelations the New City of Jerusalem is a cube. The wall of the city is 144 cubits or 216 Feet.

Revelation 21:17

He also measured its wall, 144 cubits by human measurement, which is also an angel's measurement.

Yet more proof of the UTTER PERFECTION of the numbers I found in Metatron's Cube. When I table them in 12 Rows of 18 Columns, (As the Ancients were taught to do in the earlier table) each column adds up to a perfect 60.

There is no "SIMPLE" explanation for this. This is deep programming and highly advanced Intelligence.

If anyone knows the true significance of this discovery please contact me on Facebook under Lucien Khan or on Twitter at Metatron@Fractal216

And Please share this discovery with the world. It belongs to all people. This will unite the world and all religions. The Seed of Life and Sacred Geometry is a bridge that unites all cultures.

There are still many secrets hidden inside this 60 number pattern. Do some research on the Bible Code and you will find the numbers 0,1,1,2,3,5,8.....

Here is the 216 Letter Sacred Name that was Hidden inside Metatron's Cube:

11235831459437774149325729111235831494377741561785381911235831459437774161785381999875279694377741561785381999875279651673336961785381999875279616733369549325729199875279651673336949325729111235831416733369549325729 1

This is what happens when I table the 216 Numbers found inside Metatron's Cube, into 18 columns of 12 rows.

1	2	3	4	5	6	7	8	9	10	11	12	13	14	15	16	17	18
1	1	2	3	5	8	3	1	4	5	9	4	3	7	7	7	4	1
4	9	3	2	5	7	2	9	1	1	1	2	3	5	8	3	1	4
9	4	3	7	7	7	4	1	5	6	1	7	8	5	3	8	1	9
1	1	2	3	5	8	3	1	4	5	9	4	3	7	7	7	4	1
6	1	7	8	5	3	8	1	9	9	9	8	7	5	2	7	9	6
9	4	3	7	7	7	4	1	5	6	1	7	8	5	3	8	1	9
9	9	8	7	5	2	7	9	6	5	1	6	7	3	3	3	6	9
6	1	7	8	5	3	8	1	9	9	9	8	7	5	2	7	9	6
1	6	7	3	3	3	6	9	5	4	9	3	2	5	7	2	9	1
9	9	8	7	5	2	7	9	6	5	1	6	7	3	3	3	6	9
4	9	3	2	5	7	2	9	1	1	1	2	3	5	8	3	1	4
1	6	7	3	3	3	6	9	5	4	9	3	2	5	7	2	9	1
60	60	60	60	60	60	60	60	60	60	60	60	60	60	60	60	60	60

Each Column adds up to a perfect 60. There is no simple explanation for this. This shows deep programming to the very structure of the Universe and confirms that the Universe is a SUPER INTELLIGENT design. I have found God's signature. Gods Hidden 216 letter Secret Name is hidden inside Metatron's Cube.

Diagram 9 below is a rendering of Metatron's Cube inside the Seed Of Life.

DIAGRAM 9

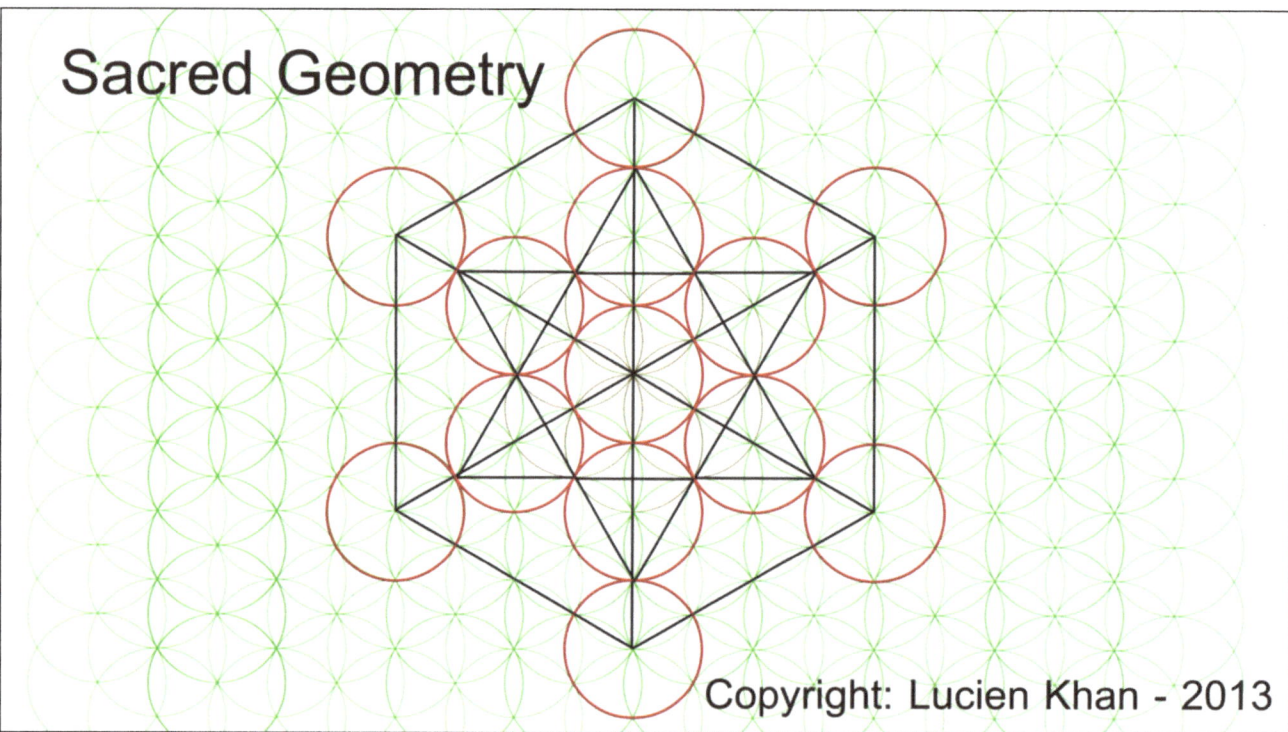

This pattern is a fractal and extends into infinity.

Go and research the numbers 54, 108 & 216. And also research who was Metatron or Enoch.

ANGELS AND ANGLES

CHAPTER 8

Angels and Angles

Look at this final rendering (Diagram 10) of Metatron's Cube highlighting the geometric angles found inside Metatron's cube.

If nothing I have said thus far has convinced you of the Divine nature of all things and the Supreme Intelligence at work, allow me to make on last effort.

DIAGRAM 10

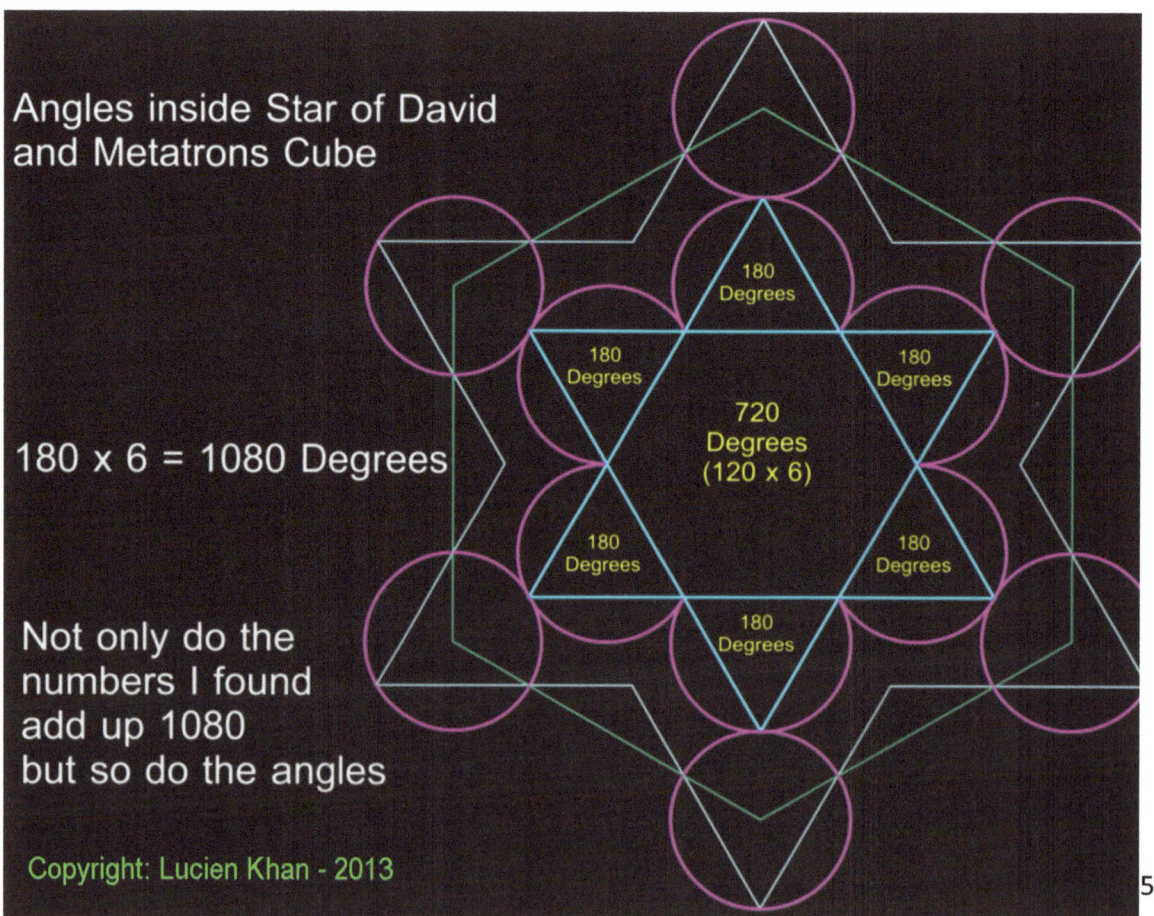

The six equilateral angles inside the Star of David that fits so perfectly into Metatron's Cube add up to exactly 1080. (6 x 180 = 1080).

If you look at my previous rendering of Metatron's Cube (**Diagram 5**) constructed using the Fibonacci 60 Digit Repeat Cycle you will see that the flower pattern around the 6 triangles also add up to 1080.

Remember our 72 names of Angels. The Hexagon in the centre of Metatron's Cube adds up to 720 degrees.

Angles have absolutely nothing to do with the 60 numbers in the Fibonacci Repeat Cycle.

And yet, there it is, clear as day, absolute proof of the sheer-mind boggling Intelligence that is GOD.

There is no way this can happen without SUPREME "UBER" ULTRA INTELLIGENCE.

GOD IS EVERYTHING.

It all starts with Accepting God.

It sounds simple, but it truly is life changing. Once you know that God is real and that every single thing that happens has purpose and meaning, your life will have new purpose and meaning.

Just try to see the world in that LIGHT.

So what is the Secret name of God?

Well, If everything builds from 108 (See image of The Seed of Life and Metatron's Cube). And Jewish Religion believed that God's secret name has 216 letters.

And I found 216 numbers that add up to 108. Then basically these 216 numbers are spelling 108.

And these 2 numbers 108 and 216 prove beyond a shadow of doubt that the Universe is not a random singularity event, as per the 'Big Bang' but a Supremely Intelligent Design.

Then I have revealed the 216 Letter name of God. Exactly as foretold.

Coincidentally my name Lucien means LIGHT and Lucien is also an anagram of the word NUCLEI.

Everything, everything that has ever happened in the Story of Mankind proves that God is the CREATOR.

WE ARE ALL THE CHILDREN OF GOD.

But why 108?

Is there any final significance to the number 108? Why does everything build from 108? Why do so many cultures think 108 is sacred? Why do the 216 numbers inside Metatron's Cube spell (add up to) 108?

It all goes back to Gematria the traditional Jewish system of assigning numerical value to a word or phrase. A number can have symbolic meaning.

Why do people meditate on the Number 108? Why does a Mala have 108 beads? Why is God believed to be 108?

Look at the number and think in terms of symbols or phrases.

This all began when I started searching for answers to what the Universe is and what God is?

My first conclusion was that the Universe must be an ETERNAL cycle of energy to mass as per Einstein's $E=MC^2$. Because Energy is eternal, it cannot be created or destroyed.

So the universe must be an ETERNAL CYCLE (Not a physical circle but a Divine Cycle of Energy to Mass). There is no permanent place called "The Universe" the Universe in only an event or story unravelling over time. The Universe only exists in time. It started at a single point in time and it will end at a single point in time. And then recycle for all eternity.

My next conclusion was that the entire Universe must be One. All connected. One entity. One event. The only thing that separates us is TIME. Watch my videos explaining this.

My final conclusion was that the entire Universe is One entity that recycles for all ETERNITY or into INFINITY.

What does 108 tell us? What is it showing us?

1 (One) 0 (Circle) 8 (Symbol for Eternity or Infinity - ∞)

The Universe is 1 Divine Circle that Recycles for all Eternity.

God is the energy that creates all things. $E=MC^2$.

Energy cannot be created or destroyed (First Law of thermodynamics).

GOD IS EVERYTHING FOR ALL ETERNITY = 108

One last thing.

I have to tell you how I came to find this 60 Digit Repeat Cycle and make all the connections that I did.

As I mentioned at the start of this book, I was working on a cosmological theory that the universe is a circle. To be honest, at the very beginning I actually thought of the universe as a physical thing, so I imagined it as a physical circle or torus shape.

Now I understand that there is no outside the universe because it is not a permanent or physical thing. It doesn't have a single centre; every single object inside the universe is at the centre of the universe. Nuclei.

This can only make sense once you can understand that the universe is only a story, an event, something that is happening or unfolding over time from the perspective of every single object.

The universe is not something you look at and say it has length, width and height. No. It is something you experience from your centre radiating outward. It is a spiritual experience.

You are at the centre of a divine universe created for you by God. I'm digressing.

How did I go from trying to prove that the universe is a divine circle to finding the answers hidden inside Metatron's Cube?

Everything around me spoke to me and guided me. When I needed inspiration I would pick up a novel and there would be an answer right there, waiting, but so profound that I knew it was left there for me by God.

It is crucial that you understand this part. Whenever you act out of goodness and purity of thought, you are doing God's work. You are leaving a marker for anyone else that looks for God.

I would find these markers left like breadcrumbs leading me all the way to God. These signs are all around you. In books, movies, games, in the very people that surround you. God is right there. All you have to do is open your eyes and accept that God is everything.

The biggest or greatest markers are the ones left by people who act out of purity of spirit. People who are inspired by God and not by corporate greed or acquisition of wealth. People who do art for art sake. People who love out of goodness and kindness and not for personal gain or selfish reward. This is where God shines the brightest.

I simply followed these signs, placing value in everything other people have said and they led me all the way to God.

Everything you do is an act of God. God is inside you. The Kingdom of Heaven is inside you.

Know this. Act upon this. And surely goodness and mercy shall follow you all the days of your life.

Thank You for your time.

I tell you now, that this book is written with pure sincerity. This is the LIGHT of God and this is a marker. This will lead you to God. You don't have to naively believe everything I'm saying, but be inspired to look for God. Life is full of magic and mystery.

Let this be the start of your journey. If you look for God you will find God. I have given you a big step forward. I have proved that the universe must be an intelligent creation. Unless you still believe all of this is random chaos.

If the universe is an intelligent design. There must be an intelligent designer.

As I mentioned at the start of this book, this 216 Letter name is based on my own interpretation or deduction, I personally cannot think of another 216 Letters or Numbers that could have as much power as these do. They point directly to an intelligent design. And I'm sure there are still many secrets waiting to be unlocked using this key.

GOD IS EVERYTHING

Remember to look out for my video's on Youtube explaining all of this in more detail. And see my online documents showing the images found in this book in a better resolution.

John 8:12

"I am the light of the world. Whoever follows me will never walk in darkness, but will have the light of life." – Jesus Christ.

Contact me at: Lucien_Khan@hotmail.com or on Twitter at Fractal216@Metatron.

Acknowledgments

Thanks to my wife Bronwyn and my daughter Erin for all their support.

And thank you to everyone who has ever been inspired to create something honest and pure, whether it is a painting, a book, a piece of architecture, a poem, a game, a song, whatever.

When you act out of love and passion you channel God and you leave a marker pointing toward the Divine.

The Tyger
William Blake

Tyger Tyger. burning bright,
In the forests of the night;
What immortal hand or eye.
Could frame thy fearful symmetry?

In what distant deeps or skies.
Burnt the fire of thine eyes?
On what wings dare he aspire?
What the hand, dare seize the fire?

And what shoulder, & what art,
Could twist the sinews of thy heart?
And when thy heart began to beat.
What dread hand? & what dread feet?

What the hammer? what the chain,
In what furnace was thy brain?
What the anvil? what dread grasp.
Dare its deadly terrors clasp?

When the stars threw down their spears
And watered heaven with their tears:
Did he smile His work to see?
Did he who made the lamb make thee?

Tyger Tyger burning bright,
In the forests of the night:
What immortal hand or eye,
Dare frame thy fearful symmetry?